Be a

MARINE BIOLOGIST

By Zelda Salt

Gareth Stevens
PUBLISHING

Please visit our website, www.garethstevens.com. For a free color catalog of all our high-quality books, call toll free 1-800-542-2595 or fax 1-877-542-2596.

Library of Congress Cataloging-in-Publication Data

Names: Salt, Zelda, author.
Title: Be a marine biologist / Zelda Salt.
Description: New York : Gareth Stevens Publishing, [2019] | Series: Be a scientist! | Includes index.
Identifiers: LCCN 2018017396| ISBN 9781538229972 (library bound) | ISBN 9781538231197 (paperback) | ISBN 9781538231258 (6 pack)
Subjects: LCSH: Marine biologists–Vocational guidance–Juvenile literature. | Marine biology–Vocational guidance–Juvenile literature.
Classification: LCC QH91.45 .S25 2019 | DDC 577.7023–dc23
LC record available at https://lccn.loc.gov/2018017396

First Edition

Published in 2019 by
Gareth Stevens Publishing
111 East 14th Street, Suite 349
New York, NY 10003

Designer: Katelyn E. Reynolds
Editor: Therese Shea

Photo credits: Cover, p. 1 Photo Researchers/Science Source/Getty Images; cover, pp. 1–32 (background image) Rik Pat/Shutterstock.com; p. 5 courtesy of NASA; p. 7 Michael Bogner/Shutterstock.com; p. 9 Imfoto/Shutterstock.com; p. 11 William Frederick Mitchell orginaly published in the Royal Navy in a series of illustrations/Ian Dunster~commonswiki/Wikipedia.org; p. 13 Benoit Daoust/Shutterstock.com; p. 15 Sirachai Arunrugstichai/Moment/Getty Images; p. 17 Mark Conlin/Oxford Scientific/Getty Images; p. 19 Stephen Frink/Image Source/Getty Images; p. 21 George Rinhart/Corbis via Getty Images; p. 23 RaksyBH/Shutterstock.com; p. 25 Richard Whitcombe/Shutterstock.com; p. 27 Keipher McKennie/WireImage/Getty Images; p. 29 Reinhard Dirscherl/WaterFrame/Getty Images.

Printed in the United States of America

CPSIA compliance information: Batch #CW19GS: For further information contact Gareth Stevens, New York, New York at 1-800-542-2595.

CONTENTS

WORDS IN THE GLOSSARY APPEAR IN **BOLD** TYPE THE FIRST TIME THEY ARE USED IN THE TEXT.

Water, Water
EVERYWHERE

The world is 71 percent water. Some of it is salty, like the ocean, and some of it is fresh, like the Great Lakes. Some of it is even a mix of both, which is known as brackish water. Chesapeake Bay in Maryland and Virginia is an example of brackish water.

Marine biologists work with plants and animals that live in salt water. Some of them might work in a lab, studying **plankton**. Or they might be found scuba diving with whales to track their movement patterns. You might even meet one at a zoo or aquarium, working as a dolphin trainer!

HOW LOW CAN YOU GO?

How deep is the ocean? You could put Mount Everest, the world's tallest mountain, upside-down in the ocean and it would reach down 29,029 feet (8,848 m)...and there would still be ocean underneath! Humans have gone even deeper, using special vessels to go further into the ocean.

MOST OF THE EARTH'S SURFACE IS COVERED IN WATER, AND 96.5 PERCENT OF THAT WATER IS SALT WATER IN THE OCEANS!

WHAT IS MARINE BIOLOGY?

Marine biologist Maddalena Bearzi sums up her field as "the study of marine organisms, their behaviors, and interactions with the **environment**." There are many ways marine biologists can work. They might study microbiology, learning about the smallest organisms on Earth. Or they might research behavior **ecology**, which is the study of what animals do and why.

OCEAN OUTLOOK

Many marine biologists study how the ocean affects the organisms that live there. They might look at how marine organisms have **adapted** to their watery environment. For example, a marine biologist might look at how an organism has adapted to living in the dark depths of the ocean.

A marine biologist might also focus on **biotechnology**, which includes learning how to copy animal adaptations for human use. One example of biotechnology is when scientists recreated the biological makeup of shark skin. They used this "shark skin" to build doorknobs that certain kinds of germs and viruses couldn't stick to. This could help stop the spread of dangerous bacteria.

SHARK SKIN HAS LOTS OF LITTLE RIDGES ON IT, WHICH MAKES IT TOUGH FOR BACTERIA TO SURVIVE.

HITTING THE BOOKS

Different marine biology jobs need different levels of schooling, but to become a marine biologist, you will need a bachelor's degree. This could be in a wide group of fields, such as ecology, molecular biology, or **zoology**. All of these fields will help prepare you for a career studying oceans.

CENTRIFUGAL FORCE

Have you ever spun yourself around as fast as you could go? Did you feel your arms lifting up, as if they were pulling away from you? That's centrifugal force, which seems to pull something away from the center of a circular path. Machines that use centrifugal force can help marine biologists in many ways, such as separating DNA from a sample.

The name of the degree may be less important than the skills you gain! A marine biologist must have the ability to think critically and solve complex problems. They also use lab equipment like centrifuges, or machines that use centrifugal force to split substances, and work with a variety of computer programs. These skills can help a budding marine biologist build a résumé and succeed in their career.

TOP US SCHOOLS FOR DEGREES IN MARINE BIOLOGY

IF YOU WANT TO BECOME A MARINE BIOLOGIST, HEAD TO THE COAST! MANY AMERICAN UNIVERSITIES NEAR THE OCEAN OFFER DEGREES IN MARINE BIOLOGY.

SCHOOL	LOCATION
Duke University	Durham, NC
Boston University	Boston, MA
University of Maine	Orono, ME
Eckerd College	St. Petersburg, FL
Oregon State University	Corvallis, OR
University of Hawaii at Hilo	Hilo, HI
University of New England	Biddeford, ME
University of New Hampshire	Durham, NH
University of North Carolina	Chapel Hill, NC
University of Tampa	Tampa, FL

Adventure COLLECTOR

Since the beginning of marine biology as a field, scientists interested in studying ocean life have had to be adventurers at heart. One of the earliest marine biologists was someone you might have heard of: Charles Darwin! Darwin journeyed around the world on a boat called the HMS *Beagle*. During his journey, he collected and studied marine organisms and other plants and animals.

A few decades after Darwin's trip, Sir C. Wyville Thomson, another early marine biologist, sailed on the HMS *Challenger* on a 3-year journey through the oceans of the world. Thomson and his team collected so much data they needed 50 volumes, or books, to write it all down!

MARINE LIFE IN NUMBERS

The **Census** of Marine Life was a 10-year project that brought scientists from more than 80 countries together to figure out what lived in the ocean. The census looked to answer three questions: 1) What has lived in the oceans? 2) What currently lives there? and 3) What will live there in the future? In the end, scientists listed and described more than 1,200 new marine species.

The *Challenger's* trip is often referred to as the birth of marine biology as a field. The information gained from the journey was studied and used for many years.

ICHTHYOLOGISTS

You may have heard of Shark Week, but do you know who completes the research on sharks? Ichthyologists study sharks, fish, and stingrays. Almost all ichthyologists have master's degrees in their field.

An ichthyologist may identify different types of fish, check water quality, or look at data. They often use their degree to further our understanding of biology, animal behavior, ecology, and environmental science.

Ichthyologists sometimes need an unusual skill! A lot of their research happens underwater, where an ichthyologist is up close and personal with the marine subjects. Many ichthyologists are scuba certified, which is when someone learns how to use special equipment to breathe underwater.

ANCIENT ICHTHYOLOGY

Humans are now much better equipped to swim with and study marine life than we ever have been before. But lack of equipment didn't stop our ancestors from making underwater discoveries. Dentists in ancient Greece used venom from a stingray's spine as an **anesthetic**. This venom can be deadly, even after the stingray dies!

ALTHOUGH STINGRAYS AND SHARKS MIGHT LOOK LIKE SOME FISH,
THERE IS ONE BIG DIFFERENCE BETWEEN THEM AND FISH.
NEITHER STINGRAYS NOR SHARKS HAVE BONES, WHILE FISH DO!

DIVE IN!

Getting a scuba certification can be a practical skill for a marine biologist. There are three steps to getting scuba certified. First, you'll complete classroom learning. Then, you will practice diving in a confined water space. Finally, you'll head out into open waters.

GET CERTIFIED!

You don't have to wait until you get your degree to get certified for scuba. Depending on where you live, you can get certified once you are 10 years old or older. Anyone under the age of 15 will earn a junior certification, which can be upgraded once they turn 15. So, suit up!

Before you put a toe in the water, you'll have to learn the basics, like what every piece of scuba equipment does, what signals divers use to talk to each other underwater, and what goes into planning a dive.

First, you'll practice using your gear in a swimming pool. Once you're comfortable, you'll go on a couple of open water dives with an instructor and complete the final test.

FOR MANY PEOPLE, A DAY AT THE OFFICE MEANS SITTING AT A DESK.
FOR A MARINE BIOLOGIST, IT MIGHT MEAN GOING UNDERWATER!

Sustainability Scientists

Another important way that marine biologists affect our daily life is through fisheries science. This is a big field that ties together all the ways that seafood gets from the ocean to your plate.

SUSTAINABLE LIVING

There is a lot of talk about sustainable living, sustainable food, and sustainable resources. But what does it mean for something to be "sustainable"? If something is sustainable, it can last for a long time, or be used without being completely used up. This is important for the protection of our world's environment.

Earth's climate has begun to change. A number of fish species are being overfished. This happens when too many adult fish are caught and the population of fish has trouble surviving. Overfishing threatens many populations of fish species, such as the bluefin tuna. Many marine biologists are focusing on how humans and sea life can live in harmony, balancing our nutritional needs with keeping our marine **ecosystems** healthy.

EAT THIS FISH, NOT THAT FISH

CERTAIN SPECIES OF FISH ARE IN DANGER OF BEING OVERFISHED. INSTEAD OF EATING THOSE SPECIES OF FISH, WE CAN INSTEAD EAT ANOTHER SPECIES THAT TASTES SIMILAR!

OVERFISHED SPECIES	SUSTAINABLE SWAP
We eat a lot of bluefin tuna.	Instead, we could eat albacore tuna.
We eat a lot of Atlantic salmon.	Instead, we could eat Alaskan wild salmon.
We eat a lot of haddock.	Instead, we could eat bib or coley.
We eat a lot of Atlantic cod.	Instead, we could eat pollack.

BLUEFIN TUNA

STUDY AND RESTORE

Marine biologists not only study the ocean, but they also can play a role in helping to restore it. Scientists are always looking at ways to undo the negative impact humans have on the environment. They are also working to stop further damage from being done.

One example is the work being done to help the Great Barrier Reef in Australia. Once one of the most **diverse** areas of the world, coral there has been dying at a high rate, in part thanks to pollution. Recently, scientists have begun a project to increase the **fertility** of corals in the Great Barrier Reef. This will hopefully bring the reef back to full health.

WHAT IS CORAL BLEACHING?

Coral is colorful because of the algae that lives in its tissue. Algae are plantlike living things that are mostly found in water. When the water around coral gets too warm, which happens because of climate change, the coral loses that algae. Without algae, coral turns completely white. If there are no algae, many corals can't feed themselves, and they starve to death.

IF SCIENTISTS DON'T FIND A WAY TO STOP CORAL BLEACHING, THE GREAT BARRIER REEF MAY DIE OFF COMPLETELY AS EARLY AS 2050.

Sometimes fighting to help and protect the oceans takes marine biologists far away from the oceans they study. Marine biologists often talk in courts about oil spills, illegal fishing, and other man-made disasters.

One well-known example is marine biologist Rachel Carson. She wrote four books in her lifetime, including three books about the ocean: *Under the Sea Wind, The Sea Around Us,* and *The Edge of the Sea.* These three books celebrated, or honored, the diverse life in the ocean. Carson's writing detailed the science of the ocean and what life is like for different creatures that live underwater. She encouraged her readers to think about how their actions affect marine life.

DDT

Rachel Carson also wrote *Silent Spring*, one of the first major publications to warn people about the danger of using pesticides, or something used to kill pests, such as bugs. Carson argued that DDT was harmful to birds. DDT and other pesticides are also very dangerous in marine environments. When they enter the ecosystem, they can sicken and kill fish and other marine animals.

IN 1952, RACHEL CARSON'S *THE SEA AROUND US* WON THE NATIONAL BOOK AWARD. CARSON'S BOOKS ARE STILL PUBLISHED AND ENJOYED BY READERS TODAY.

A Higher PORPOISE

Some marine biologists spend less time completing lab work and more time working hands-on with marine mammals. Zoos and aquariums often employ marine biologists as their animal trainers and handlers. Many trainers have a history of working with animals, such as at veterinary offices or their local zoo.

FREE WILLY?

Some mammals, such as orcas, have been shown to do poorly in captivity, which means being held in a cage. This may be because their habitat is smaller than it would be in the wild. In the wild, orcas are known to travel more than 62 miles (100 km) per day.

A bachelor's degree in marine biology, with a focus on animal behavior, is hugely helpful for animal trainers. Animal trainers may work with seals, sea lions, dolphins, or whales.

Sometimes a lot of public speaking is involved with being an animal trainer. Most trainers perform demonstrations for zoo and aquarium visitors.

MARINE MAMMAL TRAINERS ARE OFTEN STRONG SWIMMERS AND SCUBA CERTIFIED.

BIOTECHNOLOGY

Marine biologists who work in biotechnology have a unique outlook on marine life. While they study marine organisms, they focus on how the marine world can help both humanity and our planet.

For example, marine biologists specializing in biotechnology have discovered that types of algae absorb pollutant gases, like carbon dioxide. Marine biologists are even researching ways that marine organisms can be used to create pollution-free fuel.

Biotechnology is an exciting field for marine biologists. There are over 200,000 known species in the ocean and millions still undiscovered. Just think of all they have to teach us!

WIND AND WATER

Marine biologists studying humpback whales realized something interesting. They could use the structure of whale flippers to make a wind **turbine** more effective! A humpback whale's flipper has bumps along the edge. These bumps allow them to make tight turns smoothly underwater. Similar bumps were added to wind turbine blades. These new blades increased the clean energy generated, making the blades more powerful.

Although scientists believe millions of species have yet to be discovered, they also believe the total number of marine species is falling. They believe many species cannot adapt to the changes in their environment.

TECH TALK

Marine biologists use new, interesting technology. In many ways, marine biologists are at the cutting edge of technology. Marine biologists may use advanced computers and microscopes to study marine life. Some even use deep-sea robots to gain access to places deep in the ocean.

WHO TURNED OFF THE LIGHTS?

At 13,000 feet (3,962 m) below sea level, there isn't any sunlight. Here, the temperature of the water is almost at freezing. The creatures that live in this type of environment have adapted to the conditions. It's very hard to study marine animals this deep in the ocean.

Marine biologists also use manned submersibles, or vehicles that can go underwater, to go places no other humans can go. The deepest anyone has ever gone was 35,756 feet (10,898 m) deep!

However, some of the best tools are the traditional ones. Marine biologists often collect specimens, or samples of a group, using nets and buckets.

IN 2012, FILM DIRECTOR JAMES CAMERON MANNED THE DEEPSEA CHALLENGER AND WENT DOWN INTO THE CHALLENGER DEEP, THE DEEPEST POINT OF THE OCEAN THAT WE KNOW OF.

JUMP IN!

Whether you're an animal or technology lover, marine biology has something for anyone interested in the way life works in our oceans. So much of human life relies on the health of our aquatic ecosystems. We are still only at the early stages of understanding—and lessening—the impact of human activity.

BY THE NUMBERS

The ocean contains so much of life on Earth. But we still haven't developed the technology to reach every part of the ocean. We have only explored about 5 percent of the ocean as a whole! Scientists believe that over 90 percent of all marine life has yet to be discovered.

Marine biologists can start as swimmers, animal trainers, adventurers, inventors, and even **activists**. They are at the cutting edge of technology, bringing together lab research and field exploration. There's no end to the opportunities in marine biology. There is so much left to discover in our Earth's oceans!

THERE ARE MILLIONS OF SPECIES IN THE OCEAN THAT WE STILL KNOW NOTHING ABOUT. WILL YOU HELP DISCOVER THEM?

GLOSSARY

activist: one who acts strongly in support of or against an issue or cause

adapt: to change to suit conditions

anesthetic: a substance that makes a person lose feeling in a part of their body

biotechnology: the use of living things, like bacteria, to create helpful products

census: the official process of counting the number of things and collecting information about them

diverse: differing from each other

ecology: a science that looks at the interaction between living things and their environments

ecosystem: all the living things in an area

environment: the conditions that surround a living thing and affect the way it lives

fertility: having the ability to make babies

plankton: a tiny plant or animal that floats in the ocean

turbine: a motor operated by the movement of water, steam, or air

zoology: a science that looks at animals and their behavior

For More Information

Books

Gibson, Karen Bush. *Marine Biology: Cool Women Who Dive.* White River Junction, VT: Nomad Press, 2016.

McAllister, Ian and Nicholas Read. *The Great Bear Sea: Exploring the Marine Life of a Pacific Paradise.* Victoria, BC: Orca Book Publishers, 2013.

Wilsdon, Christina. *Ultimate Oceanpedia: The Most Complete Ocean Reference Ever.* Washington, DC: National Geographic Kids, 2016.

Websites

Marine Biology: The Living Oceans
www.amnh.org/explore/ology/marine-biology
Play games, hear from real marine biologists, and create coral reefs!

MarineBio
marinebio.org/
Find out everything you could ever want to know about marine biology, ocean ecology, and even marine life with the marine species database.

Professional Association of Diving Instructors
www.padi.com/
Learn more about scuba diving, sign up to get certified, and get involved with conservation efforts through PADI.

INDEX